WEATHER WATCH

Rain

CAUSES AND EFFECTS

Philip Steele

Franklin Watts

New York   London   Toronto   Sydney

Franklin Watts
387 Park Avenue South
New York, NY 10016

Library of Congress Cataloging-in-Publication Data

Steele, Philip.
  Rain / by Philip Steele
    p. cm.-- (Weather Watch)
  Includes index
  Summary: Describes the meteorological conditions under which
precipitation is most likely to occur
  ISBN 0-531-10989-5
  1. Rain and rainfall-- Juvenile literature. (1. Rain and rainfall.)
I. Title II Series: Steele. Philip. Weather watch
OC 924.7.S74 1991   90 - 414 29
551. 57'7--dc20   CIP   AC

Printed in the United Kingdom

Design: Jan Sterling
Picture researcher: Jennifer Johnson
Illustrators: Tony Kenyon, Gecko Ltd

Photograph acknowledgements

t = top, *b* = bottom

Cover: (center) Raj Kamal / Robert Harding, (outer) Geoff Renner / Robert Harding;
Title page: Maurice Nimmo / Frank Lane Picture Agency; p4*b* Robert Harding
Picture Library; 5*t* Winfried Wisniewski / Frank Lane: 6*b* L West / Frank Lane;
8*b* Robert Harding; 9t US Army / Frank Lane; 10*b* Dave Currey / NHPA; 12*b* L West
/ Frank Lane 15*b* NOAA / NESDIS / Science Photo Library; 18*b* Ernest R Manewal /
Barnabys Picture Library; 19*t* Barry Waddams; 20*b* Patrick Fagot / NHPA; 21*b*
Franz J Camenzind / Planet Earth Pictures; 22*b* Anthony Bannister / NHPA; 23*t* T
Ives / ZEFA Picture Library; 24*b* David T Grewcock / Frank Lane; 25*b* Ivan Polunin /
NHPA; 26*t* Freeze Frame / Quadrant Picture Library; 27*b* Robert Harding; 28*b*
Lowell Georgia / Science Photo Library.

Contents

Here comes the rain!

People, animals and plants need **water** to live. People can live for quite a long time without food, but only a short time without water. So rain is important to everyone on Earth, filling lakes and rivers with life-giving water and helping plants to grow.

In India, from the beginning of March, the days are hot and dry, without a **cloud** in the sky. Then around the end of June, when the heat is at its greatest, the rain comes. Huge sheets of water fall to earth, for weeks and weeks on end without stopping. Dried-up riverbeds become rushing torrents, and plants come to life once more.

In other parts of the world, it can rain at any time. For example, people who live near the sea, with mountains behind them, usually have rain all the year around. Often just on the other side of the mountains, little rain may fall. The ground dries and cracks, turning it into a **desert** where few plants and animals can live.

The monsoon rains are welcomed in India even if they cause flooding in the cities. The ending of a drought, a long period without rain, means that the crops will grow and that there will be enough food.

Water brings life to our planet. It is in the air, in our bodies, and in the plants all around us. Without rainfall the earth would be a desert.

The watery planet

From space, a great deal of the planet Earth appears blue, because over 70 percent of it is covered with water. White clouds shield both sea and land from the harsh rays of the sun. The clouds are made up of tiny droplets of water condensed from **water vapor**. This is one of the invisible gases that make up the air we breathe.

Rain forms when the water vapor turns from a gas into a liquid. This is part of what is called the **water cycle** or hydrologic cycle.

The water cycle

At the start of the cycle, water from the surface of the land and oceans **evaporates** in warm air, turning into invisible vapor. The warm air rises and expands, taking up more space. The higher the air rises, the colder it becomes. This makes the water vapor in it condense into droplets.

The water can now be seen again, as it forms clouds. The droplets join together, and as soon as they become big enough, they fall as rain. When the rain falls, it fills the lakes, rivers and oceans with water. This water evaporates in its turn. It rises, forms clouds and once more returns to the ground as rain, to start the cycle all over again.

Sometimes, if the droplets in the clouds are very cold, they may freeze and fall as **hailstones**. Water vapor in the clouds can also form ice crystals, which stick together to create snowflakes. However, clouds do not always produce rain, hail or snow. If they meet warm air, they turn back into water vapor.

In **meteorology**, the science of weather, the process which brings about rain, hail or snow is called **precipitation**.

A single drop of rain hanging from a twig. When the drop breaks, the water will flow away into the ground.

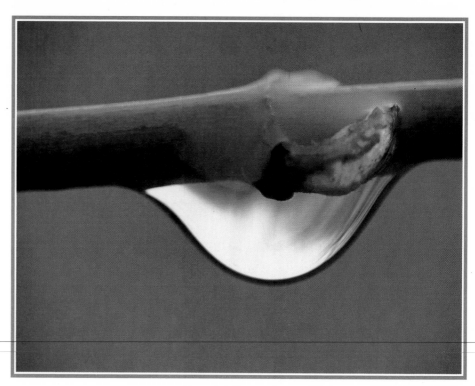

Our planet's waterworks keep running day and night. Rivers, oceans, clouds and rainfall are all part of an endless water cycle.

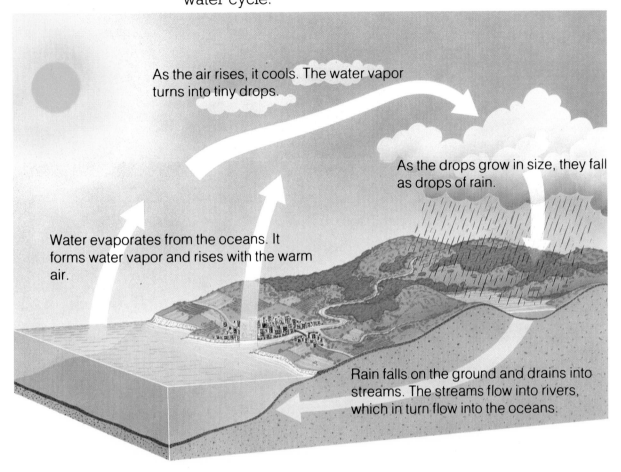

As the air rises, it cools. The water vapor turns into tiny drops.

As the drops grow in size, they fall as drops of rain.

Water evaporates from the oceans. It forms water vapor and rises with the warm air.

Rain falls on the ground and drains into streams. The streams flow into rivers, which in turn flow into the oceans.

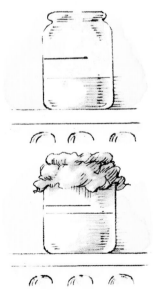

Find out about evaporation

Pour water into two jars (the same shape and size) so that each jar is half full of water. Make sure the water level is the same in both, and mark the level on the outside. Put a foil cover on one jar, then leave both jars in a warm place for several days.

Now, which jar has less water in it?

Although heat has made the water evaporate in both jars, the one with the cover on it has more water in it. This is because the foil cover has stopped the water vapor escaping into the air.

What is rain?

Rain is water that falls from the sky in liquid droplets. Water itself is a colorless liquid with no taste or smell, that freezes at a temperature of 0°C (32°F) and boils at 100°C (212°F). Although two gases, **oxygen** and **hydrogen**, combine to form water, by weight, there is nearly nine times more oxygen than hydrogen in it.

Rainwater, river or seawater is never pure in its natural state. It often contains gases, minerals such as salt, and specks of dust or sand.

The type of cloud and the general conditions in the air may produce various kinds of rain. A shower is a fall of rain that only lasts a few minutes. **Drizzle** is a fine, steady rain. Each drop may be as small as 0.25 mm (0.01 in) across, and fall at a speed of only 3 kph (2 mph).

The dark rain clouds gather, covering the sun. In the distance rain pours down into the sea.

A downpour or **cloudburst** on the other hand is a heavy fall of rain. Its large drops can measure up to 5 mm (0.20 in) across and fall at a speed of 25 kph (15 mph).

The swirling waters of a flash flood can wash away soft soil, and cause damage to property.

Flood and drought

Too much rain can cause rivers to flood. In some places this has helped people. When the floods go down, the river valley is filled with thick, rich mud, or silt, which is excellent for growing crops. Many farming villages came into being along the banks of rivers such as the Nile.

Floods can also be very dangerous. The Huang He, or Yellow River, is known as "China's sorrow," because it has often caused serious floods. In 1987, about 900,000 people were drowned.

Very heavy rain over a short period can cause a flash flood. Such a flood may overload city drains, creating a health hazard. A flash flood can also cause landslides or mudslides, endangering people and their homes.

Too little rain can be as bad as too much. The biggest desert in the world, the Sahara in Africa, is growing larger year by year partly because of the lack of rain.

Around the world

The weather in a particular country or place usually has a pattern over the years, and this is called its **climate**. Some places have a very wet climate, and others a dry one. Driest of all are the deserts. Many deserts are very hot, although there are some cold ones. Few plants can grow in either type of desert.

Mild or temperate countries have four seasons: spring, summer, autumn and winter. In these countries it may rain at any time.

Some parts of the world have a dry season and a wet season. During the wet season, ocean winds called **monsoons** bring heavy rains.

In some tropical countries where it is always hot and wet, the heavy **rainfall** has caused dense forests to grow. These **rain forests** are very useful to us. The trees take in large amounts of a gas called **carbon dioxide,** and give out life-giving oxygen.

Lands beside the sea or surrounding large lakes often have a mild, rainy climate. Lands in the middle of large continents often tend to have a dry climate, cold in winter and hot in summer.

Mountain ranges run down the western coast of North America. Winds from the Pacific Ocean bring rain and snow to the peaks. However, the land on the eastern side of the mountains often receives very little rain. It is called a "rain-shadow" desert. Death Valley, California, is such a rain-shadow desert; it is dry and very hot.

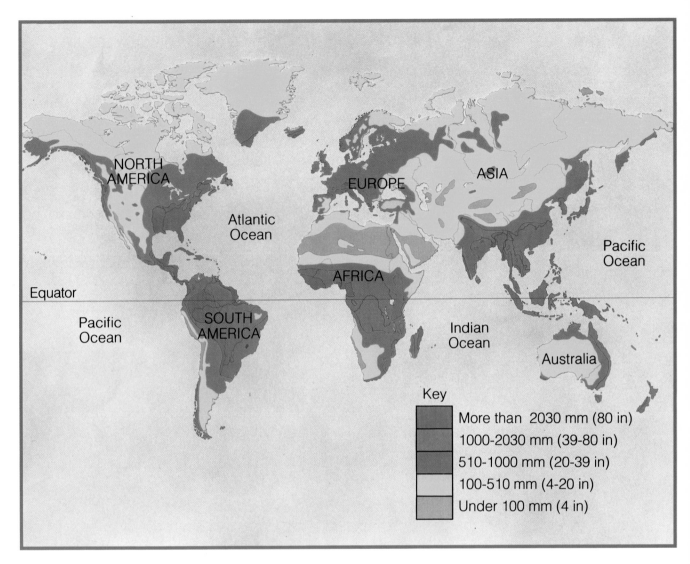

The map shows the average amounts of rain that fall each year around the world. Most areas near the equator are wet and hot. The blue- and purple-shaded areas near the equator have rain all year round. The green areas in the tropics have one wet and one dry season. Yellow areas are dry. Orange areas are deserts.

Key

	More than 2030 mm (80 in)
	1000-2030 mm (39-80 in)
	510-1000 mm (20-39 in)
	100-510 mm (4-20 in)
	Under 100 mm (4 in)

Rain belts

What makes some parts of the world wet, and others dry? Coastlines, mountains, and ocean currents all play their part. So too does the movement of large masses of air above the earth.

Air from the ocean often brings wet weather, and air from the middle of the continent brings dry weather. Warm damp air meeting cold air from the middle of a continent brings wet weather. Warm damp air meeting cold air from the Poles can sometimes produce a belt of rain 500 km (about 300 mi) wide and as long as 1,000 km (about 600 mi).

Some air masses stay still, like the hot wet ones over the rain forest regions. In places like the monsoon lands of India, the air masses take turns to produce rain at regular seasons.

Is it going to rain?

Some old sayings

Red sky in the morning, sailor take warning;
Red sky at night, sailor's delight.

Rain before seven, fine before eleven.

If cows lie down, it's going to rain.

If it rains on St. Swithun's Day, July 15th, it will rain for the next 40 days.

Rain has always been important to people. Without rain, crops might fail, and if that happens, people can go hungry. So for thousands of years, people have watched the skies for signs of rainy weather. There have always been many sayings to help people remember what the signs meant.

Night or day, the sky offers many clues. A ring of light around the moon means that rain is on the way. If the horizon is hazy, you can expect fine weather. If the horizon looks hard and clear, it will be wet.

These cumulo-nimbus clouds mean rain is on the way. Cumulo-nimbus means "heaped rain cloud."

12

Tell-tale clouds

Cloud shapes can give us the clearest warnings of rain, and there are many different types. Cumulus clouds are puffy and white. Cirro-cumulus clouds are very high. Their dappled or rippled appearance, like the markings on a fish, has given them the name "mackerel sky." Cirrus clouds are known as "mare's tails," because they are long and wispy.

Towering columns of clouds are called cumulo-nimbus, and they generally mean rain. Nimbo-stratus, which forms a low, dark curtain, also means rain.

Check the wind

Look out for a weather vane on the roof of a building in your town. If you see cumulo-nimbus or nimbo-stratus clouds building up, note which way the wind is blowing. This will tell you which winds in your area bring rain.

Make a wet and dry thermometer

The amount of water vapor in the air we breathe is called the air's **humidity**. The air is more humid when it is nearly ready to rain, and you can use this to help you to forecast.

A wet and dry bulb thermometer can tell you how humid the air is, and you can make one for yourself with two thermometers.

Tie a small wet cotton sock onto one of the thermometers, to turn it into a wet bulb thermometer. If the air is very dry, water evaporates quickly and the wet bulb thermometer shows a lower reading than the dry one. If the air is humid, however, and it is about to rain, the two thermometers read almost the same.

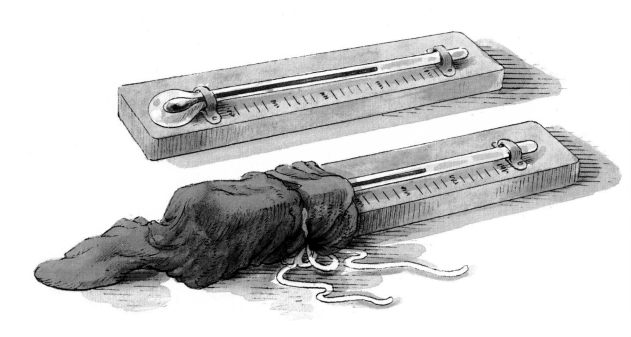

Forecasts and records

Tomorrow's weather is more important to some people than it is to others. Farmers need to harvest their crops before they are spoiled by the rain. Sailors and pilots may alter their plans if bad weather is forecast.

Today forecasters are much more accurate than they were in the past. They have a whole range of scientific equipment to help them, and also the weather satellites. These circle the earth out in space, transmitting photographs of the changing cloud patterns back to the forecasters.

This satellite picture shows rain clouds building up over the North Atlantic. Meteorologists follow the progress of the clouds to help them forecast tomorrow's weather.

Records are an important part of forecasting. Weather stations, aircraft, and even balloons are used to take measurements of the amount of humidity in the air, temperature, hours of sunshine, strength and direction of wind and rainfall. They also estimate cloud heights, visibility and cloud types.

Many different instruments are used in taking these measurements, and they must be read at least every three or four hours to give a complete record of the weather conditions over a period. One instrument used is a **rain gauge** to measure rainfall, and you can make one yourself quite easily.

A simple rain gauge

Cut the top off a plastic soft drink bottle, and fit it upside down in the bottom half of the bottle to make a funnel. The amount of water caught in the bottle will be the amount of rainfall.

Use a waterproof felt-tip pen to mark a scale on the side of the bottle: 5 mm, 10 mm, 15 mm (0.25 in, 0.50 in, 0.75 in) and so on. Your rain gauge is now ready to use.

Put your gauge in a shallow hole so that it cannot be blown over, on level open ground away from buildings or trees. Measure the rainfall over 24 hours, and then empty the bottle. Keep a record of your readings in your rain diary.

Weather maps

All the measurements from the weather stations are passed to a central station. There meteorologists make up weather maps using these records and also the information relayed from the satellites. The new maps are compared with previous maps to assess how the weather systems have changed. Then a prediction can be made for the next day's weather.

The familiar TV weather charts usually show cold and warm **fronts**, marking the borders between cold and warm streams of air. These tell us when and where it is likely to rain. **Isobars** are the other lines drawn on these much simplified charts. The curves look like contours on a map and are used to show places where the **air pressure** is the same.

A warm front produces rain and snow. A cold front produces showers and heavy rain. An occluded front occurs where a cold front overtakes a warm front and produces a bank of cloud and rain.

A barometer is an instrument used to measure air pressure. Again this helps with forecasting. When the barometer shows high pressure we will have settled weather. When the pressure is low, it means a **depression** (storm or low-pressure system) is moving in, which usually brings rainy weather.

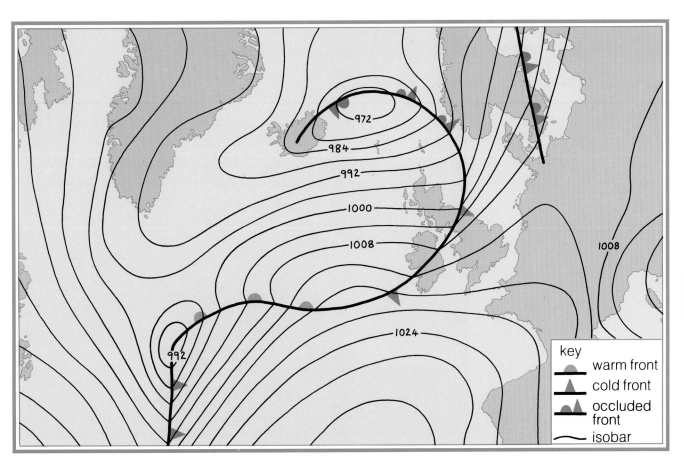

972
984
992
1000
1008
1008
1024
992

key
warm front
cold front
occluded front
isobar

Rain and landscape

The surface of our planet has been shaped by rainwater over the ages. Flowing water has gouged out deep valleys. Waterfalls have cut deep clefts in the sides of mountains. Hard rock can stand up to the water, but soft rock is worn away more quickly. Canyons, caves and natural arches are formed. The wearing down of rock by water (or wind) is called **erosion.**

The Iguazu Falls in Argentina, South America. The force of the falling water wears away at the soft rock, leaving boulders strewn at the foot of the falls.

Rain clouds forming over the mountain peaks.

Just as rain can affect the landscape, so the landscape can affect and control where and how much rain falls. For example, high mountain peaks are very cold compared with the valleys. Since warm air rises carrying water vapor with it, the mountain peaks are therefore usually covered with clouds. More rain and snow fall in highland areas than in the lowlands.

Oceans, seas and larger lakes can have a great effect on the amount of rain that falls, and so do the winds. The winds cause changes in temperature by blowing clouds up and down over the earth's surface. Those temperature changes can make all the difference as to whether we have a bright dry day or a wet one.

Warm air flows from the tropics and meets the cold air from around the Poles. In the same way, warm ocean currents meet cold ocean currents. The temperature changes caused by these movements affect the pattern of rainfall around the world.

Rain and plants

In order to survive, most plants must take in a supply of water. Up to three-quarters of a plant may be made up of water, and this keeps it stiff and fresh. Without water, the plant soon wilts.

Plants take in water through their roots, which are covered in tiny hairs. The water also contains minerals from the soil which the plant needs to make food. Some roots burrow down to reach water deep in the soil, others spread widely to soak up water near the surface of the soil. Many plants grow in ponds and rivers.

Not all rain forests are in tropical countries. The climate in Washington State is wet but cool. Moisture-loving ferns and mosses thrive amid the green trees of the Olympic National Park.

Wet and dry

Plants have developed in different ways to suit different climates. Since mosses do not have roots like ordinary plants, they take in water through their leaves. Most mosses cannot survive in dry weather, and are found in damp places such as rain forests. Ferns also usually prefer moist, shady ground.

Some plants need a very large amount of rain water. Rice is one of these, and it is a typical plant of the monsoon regions of Asia. In the wild, it grows in the flood plains of large rivers such as the Ganges. When people grow it as a crop, they have to copy these conditions by flooding the fields for the young plants to grow. This is done by **irrigating** the land with water from a nearby river.

In Asia, farmers flood their paddy fields during the monsoons so that they can plant out the young rice seedlings. When the plants are ready to harvest, the fields are drained.

Surprisingly, some plants have learned to live in deserts, where there is hardly any rainfall. Cacti store water in their spiny stems. Succulents store it in their tough leaves. Desert plants seem dusty and deadlooking much of the time. However, on the rare occasions when it does rain, the desert is suddenly bright with colorful flowers and green shoots everywhere as dry seeds sprout.

Animals and humans

Frogs have moist skins and need to remain in damp places. They are amphibians, creatures which have learned to live partly on land and partly in the water. South African rain frogs, like this one, measure 4 cm (1.5 in) in length.

Just like plants, animals need water to survive. They have to find a river, a pool of rainwater or a waterhole where they can drink. Some animals, such as the hippopotamus, live in the water most of the time. Others, such as newts and toads, return to the water to lay their eggs. Some frogs and fish, such as the climbing perch of southern Asia, bury themselves in mud during dry weather. There they can survive for long periods until the rain falls once more.

Without water, there would be no crops. Without crops, people and animals would have nothing to eat. This man is making sure that the water drains into the field properly, so his crops get enough water.

Many animals love rainy weather. Slugs and worms come out after a shower, but hide under stones and soil during the heat of the day. Woodlice always head for dark and damp places, away from direct sunlight.

Birds use oil to make their feathers waterproof so that they can fly in rainy weather. Most birds have a small gland in their bodies that produces oil. They use their beaks to spread the oil over their feathers to make them waterproof.

The human animal also needs water to stay alive. Humans cannot drink salty seawater. They need fresh water, which is taken from rivers, lakes or wells, or collected directly as rainfall.

When the rain turns sour

The lakes in North America and Europe once sparkled with health. Now fish are dying in the thousands in the lakes of the eastern United States and eastern Canada and in Scandinavia, Germany, Wales and Scotland. Pine trees are losing their needles. Other living things are being poisoned by **acid rain.**

Acid rain is caused when coal-fired power stations and factory chimneys give out poisonous smoke. The gases are carried long distances by the wind. They mix with the water vapor in the air and make it poisonous. Sometimes the raindrops that fall are more acid than vinegar!

Rain will not be clean and healthy again until the air itself is clean once more. The only way to do this is to make sure that the power stations and factories control sending out poisonous fumes.

The poisonous gases produced from the factory chimneys in one country often result in acid rain falling in another.

Living with rain

Rain affects the human world in many different ways. Land has to be drained and ditches must be dug to take away the excess water. Roads have to be covered with hard material and must also be slightly banked, or crowned, so that rainwater runs off easily. Roads that are badly built can be washed away during heavy rains.

Houses must be properly roofed so that the rain cannot come in. They also need gutters and drainpipes to take heavy rain away. The walls have to be protected by a moisture-proof layer of material. This prevents damp rising from the ground into the house. Damp buildings are unhealthy to live in.

Some rivers, such as the Mississippi, have embankments, or levees, built alongside to hold back the floods. Bridges have to be strong enough to withstand the force of floodwater swirling around their supports. In cities, storm drains are built to channel away the water left by heavy rains.

These houses in this Malaysian fishing village have been built on stilts. When monsoons cause the river to flood, the houses will remain high and dry.

Rainwater on the track causes a racing car to hydroplane and skid. Water reduces the friction between tire and track.

Rain on the road

As a car drives along, its tires rub against the road surface. This causes friction, helping the tires to grip the road. However, when it rains, the water makes the surface smooth. The tires lose their grip and begin to "hydroplane." The car may skid.

Pedestrians should take care when they are crossing roads in wet weather. The visibility is not so good in rainy weather so it is a good idea to wear bright clothes. Also because of the loss of friction cars may not be able to stop in time to avoid a pedestrian.

Friction will be increased if the road surface has a roughened texture, and if the tires have a deep, patterned tread.

Finding out about friction

Test the friction of various metals by rubbing them together. Try: rubber against stone; metal against stone; wood against plastic; and chalk against slate.

Now put the base material in a sink or dish, and cover it with a thin layer of water and try again. Does the water smooth out the friction? Does it cool down the metal?

Using rainwater

Humans have made use of rainwater for thousands of years. It has been channeled from rivers and other sources in order to irrigate crops. A number of systems were invented in ancient times to move the water into the channels, such as an arrangement of buckets and wheels operated by oxen. Some of these systems are still in use, although nowadays pumps have usually taken their place. The water is piped to the fields, then either dripped onto the roots of the crops or sprayed over a wide area. Modern irrigation methods make it possible to grow crops in desert areas. In other areas two crops can be grown during the growing season instead of just one.

Artesian wells

In some places the rock is **porous**, and soaks up water. Rainwater may sink through several layers of porous rock before meeting a layer of hard or nonporous rock. There it is trapped in a pocket beneath the ground. The level of this ground water, called the **water table**, rises and falls according to the weather. People bore holes through the rock to reach the water, which is then released. Bore holes of one special kind are called Artesian wells, because they were first dug in the Artois region of France.

Irrigation has made parts of the Sahara Desert green. Without the well in the foreground no crops could be grown in this area. Irrigation is expensive, and care must be taken that thin desert soils are not washed away.

Taming the flood

As rainwater drains into rivers, it can be used to provide power. Water power can be used to run machines. Many years ago, waterwheels were used to turn millstones to grind corn into flour. Today, the rushing water can be used to drive the blades inside **turbines**. Turbines drive machines to make electricity. Power generated in this way is called **hydroelectric**.

The Hoover Dam across the Colorado River. The hydroelectric power plant generates 1,000,000 kw from its 18 generators.

Rainmaking

Nowadays, scientists can often make it rain. Planes drop chemicals on the clouds. Moisture collects around the crystals and forms particles or water droplets. Some of the droplets join together and may fall as rain. This is called **cloud-seeding**, but so far it has not been as successful as everyone had hoped.

As technology expands, we learn more and more each day about the world we live in. Although we are a long way from being able to control the weather, we must do something to help our climate. Alternatives to fossil fuels should be used in our power stations, and levels of carbon dioxide in the atmosphere should be reduced. Otherwise rainfall patterns in some parts of the world may be affected, causing serious flooding and drought in some areas.

Keep a rain diary

Make a note each day of how much rain falls, and when. (Your rain gauge will help with this.) Note down whether it is a shower, a drizzle or a downpour.

Rain diary									
	Monday			Tuesday			Wednesday		
	8.00 a.m.	Noon	4.00 p.m.	8.00 a.m.	Noon	4.00 p.m.	8.00 a.m.	Noon	4.00 p.m.

Type of rain
  Showers
  Drizzle
  Downpour
	✓ (Drizzle)	✓ (Showers)	✓ (Downpour)						

Rain gauge reading	←——————— 50mm (2 in) ———————→								

Wind
  Light
  Medium
  Strong
	✓ (Light)	✓ (Strong)	✓ (Medium)	✓ (Medium)	✓ (Medium)	✓ (Medium)			

Clouds Type	✓	✓	✓	cumulus	cumulus	cumulus			
Sunshine				✓	✓	✓			
Thunder Lightning									
Rainbow									

Glossary

acid rain   Rain that forms from water vapor that has combined with chemicals such as sulfur dioxide. The rain releases poisonous substances when it is soaked up by the soil.

air pressure   The force with which the air presses down on the earth's surface.

carbon dioxide   A gas found in the air we breathe in and out.

climate   The average weather conditions in a region.

cloud   A cluster of water droplets or ice particles which can be seen hanging in the air.

cloudburst   A sudden, very heavy fall of rain.

cloud-seeding   The spraying of clouds by an aircraft. Chemicals such as silver iodide encourage raindrops to form. This process can be used to increase rainfall.

depression   An area of low air pressure, normally circled by rotating winds.

desert   An area which has turned into a wilderness because of very low rainfall.

drizzle   A fine, slow rain.

erosion (weathering)   The process by which rock and soil are worn away by wind and water.

evaporate   To turn into vapor.

front   The line along which one stream of air displaces another.

hailstone   A frozen raindrop, which falls as a hard ball of ice.

humidity   The amount of moisture found in the air.

hydroelectric   To do with the generation of electricity by waterpower.

hydrogen   The lightest gas known, found in the sun. Hydrogen combines with oxygen to give us water.

irrigate   To bring water to a dry area so that crops may be grown.

isobar   A line drawn on a map linking together places where the air pressure is equal.

meteorology   The scientific study of weather conditions in the air which surrounds our planet.

monsoon   A seasonal wind bringing heavy rains. The term normally applies to India and Southeast Asia.

oxygen   A life-giving gas found in the air we breathe. It combines with hydrogen to give us water.

porous   Full of tiny holes, and so able to soak up liquids. Chalk is a porous rock. Granite is nonporous.

precipitation   The process by which water vapor turns into rain, sleet, snow or hail.

rainfall   The amount of rain measured in one place over a certain period.

rain forest   A jungle or forest that grows in a moist region with a high rainfall.

rain gauge   Any instrument used to measure rainfall.

turbine   A wheel which has many curved blades. It can be spun around rapidly by a fast-flowing liquid.

water   A colorless liquid which has no smell or taste. It can also exist as a solid in the form of ice, or as a gas in the form of vapor.

water cycle (hydrologic cycle)   The endless process in which rain falls, turns into vapor, rises, turns into droplets, and falls back to earth.

water table   The level below which the soil is saturated with water.

water vapor   An invisible gas in the air we breathe. Water becomes vapor when it dries out, or evaporates.

Index